Data Analysis Workbook

Data Analysis Workbook

Starr Evers Starr

BROOKS/COLE
CENGAGE Learning™

Australia • Brazil • Japan • Korea • Mexico • Singapore • Spain • United Kingdom • United States

BROOKS/COLE
CENGAGE Learning™

Data Analysis Workbook
Starr, Evers, Starr

Publisher: Yolanda Cossio

Senior Acquisitions Editor: Peggy Williams

Assistant Editor: Elizabeth Momb

Editorial Assistant: Alexis Glubka

Technology Project Manager: Kristina Razmara

Marketing Manager: Stacy Best

Marketing Assistant: Elizabeth Wong

Marketing Communications Manager: Linda Yip

Project Manager, Editorial Production: Andy Marinkovich and Michelle Cole

Creative Director: Rob Hugel

Art Director: John Walker

Print Buyer: Linda Hsu

Production Service: Lachina Publishing Services, Inc.

Text Designer: Brian Salisbury

Photo Researcher: Linda Sykes

Copy Editor: Mandy Hetrick, Lachina Publishing Services, Inc.

Cover Designer: Brian Salisbury

Cover Image: Steve Winter/Getty Images

Compositor: Lachina Publishing Services, Inc.

Library of Congress Control Number: 2008941155

ISBN-13: 978-0-495-82894-5

ISBN-10: 0-495-82894-7

Brooks/Cole
10 Davis Drive
Belmont, CA 94002-3098
USA

Cengage Learning is a leading provider of customized learning solutions with office locations around the globe, including Singapore, the United Kingdom, Australia, Mexico, Brazil, and Japan. Locate your local office at: **international.cengage.com/region**.

Cengage Learning products are represented in Canada by Nelson Education, Ltd.

For your course and learning solutions, visit **academic.cengage.com**.

Purchase any of our products at your local college store or at our preferred online store **www.ichapters.com**.

Printed in the United States of America
1 2 3 4 5 6 7 12 11 10 09 08

CONTENTS

Peacock Butterfly Predator Defenses 1

Element Abundances in Humans, Earth, and Seawater 2

Mensink–Katan Study: Effects of Dietary Fats on Lipoprotein Levels 3

Kartagener Syndrome Phenotype 4

Bannykh Study: CFTR Distribution in Organelles 5

Pittler Study: Effects of Artichoke Extract on Hangovers 6

Tilman Study: Energy Efficiency of Biofuels from Corn, Soy, and Prairie Grasses. 7

Kuruvilla Study: Mitochondrial Abnormalities in TF Patients 8

HeLa Cell Chromosome Number 9

Case Western University Study: Meiotic Abnormalities Linked to BPA Exposure. 10

Skin Color of Indigenous Peoples Correlated to Regional Incident UV Radiation 11

Coryell Study: Genetic Relationship between Mood Disorders and Intelligence 12

Hershey–Chase Experiment. 13

Serra Study: Mutations in Individual Neurofibromatosis Tumors . . . 14

BRCA Mutations in Women Diagnosed with Breast Cancer 15

Tabuchi Study: Enhanced Spatial Learning Ability in Mice with Autism Mutation 16

Alvarez Study: Abundance of Iridium in the K–T Boundary 17

Kohn Study: Resistance to Rodenticides in Wild Rat Populations 18

Fleischer Study: Honeycreeper Phylogeny 19

Changes in Conditions on Early Earth 20

HIV Transmission from Mother to Child 21

Jacob Koella Study: Attractiveness of *Plasmodium*-Infected Hosts to Mosquitoes 22

UNFAO Report: On Deforestation 23

Effect of Logging Practices on Tree Loss to a Fungal Pathogen 24

Survival of Horseshoe Crabs after Bleeding 25

Trends in Primate Evolution 26

Heat Index (Apparent Temperature) 27

Grissino-Mayer Study: Dendroclimatology Data from El Malpais National Monument 28

Doty Study: Usefulness of Cytochrome P450 Transgenic Plants for Phytoremediation of TCE 29

Search for Pollinators of *M. depressa* Plants 30

Delphia Study: Predator-Specific Volatile Compounds Produced by Tobacco Plants 31

Effect of Cultured Skin Treatment on Diabetic Foot Ulcers 32

Jack Lipton Study: Behavioral Effect of Prenatal Exposure to MDMA . . . 33

Effects of Age and Occupational Noise Exposure . . . 34

Shanna Swann Study: Effect of Agricultural Water Contamination on Humans . . . 35

Clinical Trial of Treatment for Osteogenesis Imperfecta . . . 36

Distribution of Stroke Deaths across the United States . . . 37

Khan Study: Cervical Cancer Incidence in HPV-Positive Women . . . 38

Risk of Cancer from Radon Exposure . . . 39

George Perry Study: AMY-1 Copies among Cultural Groups . . . 40

Chensheng Lu Study: Effect of Organic Food on Pesticide Residue . . . 41

Drug Treatment to Prevent Masculinization in Females . . . 42

Multiple Births and Birth Defects . . . 43

Spread of Africanized Honeybees in the United States . . . 44

Martin Wikelski Study: Marine Iguana Populations in the Galapagos . . . 45

Methods of Controlling Fire Ant Populations . . . 46

Changes in Atmospheric Carbon Dioxide Levels . . . 47

Sea Surface Temperature Anomalies (El Niño and La Niña) . . . 48

Biological Magnification in the Arctic . . . 49

PREFACE

We are bombarded by facts and figures everywhere we go. But what does it all mean? It is important for you to be able to make sense of the information your instructors want you to understand, as well as the data you are presented with on Web sites, newspapers, and television newscasts.

This workbook gives you dozens of exercises that present real data you might encounter in your study of biology. These exercises will give you the practice you need to be better able to interpret data and build the critical thinking skills you need in and out of the classroom.

Peacock Butterfly Predator Defenses

The photographs to the right represent experimental and control groups used in a peacock butterfly experiment.

See if you can identify each experimental group, and match it with the relevant control group(s). *Hint:* Identify which variable is being tested in each group (each variable has a control).

a Wing spots painted out

b Wing spots visible; wings silenced

c Wing spots painted out; wings silenced

d Wings painted but spots visible

e Wings cut but not silenced

f Wings painted but spots visible; wings cut but not silenced

Element Abundances in Humans, Earth, and Seawater

Living and nonliving things have the same kinds of atoms joined together as molecules, but those molecules differ in their proportions of elements and in how the atoms of those elements are arranged. The three charts to the right compare the proportions of some elements in the human body, Earth's crust, and seawater.

1. Which is the most abundant element in dirt? In the human body? In seawater?

2. What percentage of seawater is oxygen? Hydrogen? How many atoms of hydrogen are there for each atom of oxygen in seawater? In which molecule are hydrogen and oxygen found in that exact proportion?

3. How many atoms of chlorine are there for every atom of sodium in seawater? What common molecule has one atom of chlorine for every atom of sodium?

Human		Earth		Seawater	
Hydrogen	62.0%	Hydrogen	3.1%	Hydrogen	66.0%
Oxygen	24.0	Oxygen	60.0	Oxygen	33.0
Carbon	12.0	Carbon	0.3	Carbon	< 0.1
Nitrogen	1.2	Nitrogen	< 0.1	Nitrogen	< 0.1
Phosphorus	0.2	Phosphorus	< 0.1	Phosphorus	< 0.1
Calcium	0.2	Calcium	2.6	Calcium	< 0.1
Sodium	< 0.1	Sodium	< 0.1	Sodium	0.3
Potassium	< 0.1	Potassium	0.8	Potassium	< 0.1
Chlorine	< 0.1	Chlorine	< 0.1	Chlorine	0.3

Comparison of the abundance of some elements in a human, Earth's crust, and typical seawater. Each number is the percent of the total number of atoms in each source. For instance, 120 of every 1,000 atoms in a human body are carbon, compared with only 3 carbon atoms in every 1,000 atoms of dirt.

Mensink–Katan Study: Effects of Dietary Fats on Lipoprotein Levels

Cholesterol does not dissolve in blood, so it is carried through the bloodstream by lipid–protein aggregates called lipoproteins. Lipoproteins vary in structure. Low-density lipoprotein (LDL) carries cholesterol to body tissues such as artery walls, where it can form health-endangering deposits. LDL is often called "bad" cholesterol. High-density lipoprotein (HDL) carries cholesterol away from tissues to the liver for disposal; it is often called "good" cholesterol.

In 1990, R.P. Mensink and M.B. Katan published a study that tested the effects of different dietary fats on blood lipoprotein levels. Their results are shown in the chart to the right.

	Main Dietary Fats			
	cis-fatty acids	*trans*-fatty acids	saturated fats	optimal level
LDL	103	117	121	<100
HDL	55	48	55	>40
ratio	1.87	2.43	2.2	<2

Effect of diet on lipoprotein levels. Researchers placed 59 men and women on a diet in which 10% of their daily energy intake consisted of *cis*-fatty acids, *trans*-fatty acids, or saturated fats. Blood LDL and HDL levels were measured after 3 weeks on the diet; averaged results are shown in mg/dL (milligrams per deciliter). All subjects were tested on each of the diets. The ratio of LDL to HDL is also shown.

1. In which group was the level of LDL ("bad" cholesterol) highest?

2. In which group was the level of HDL ("good" cholesterol) lowest?

3. An elevated risk of heart disease has been correlated with increasing LDL-to-HDL ratios. In which group was the LDL:HDL ratio highest? Rank the three diets according to their potential effect on cardiovascular health.

Kartagener Syndrome Phenotype

An abnormal form of the motor protein dynein causes Kartagener syndrome, a genetic disorder characterized by chronic sinus and lung infections. Biofilms form in the thick mucus that collects in the airways, and the resulting bacterial activities and inflammation damage tissues.

Affected men can produce sperm but are infertile. Some have become fathers after a doctor injects their sperm cells directly into eggs. Review the figure to the right, then explain how abnormal dynein could cause the observed effects.

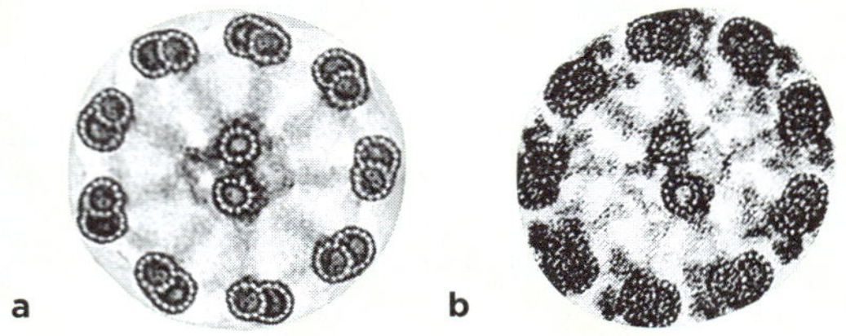

Cross-section of the flagellum of a sperm cell from (**a**) a human male affected by Kartagener syndrome and (**b**) an unaffected male.

Bannykh Study: CFTR Distribution in Organelles

In most individuals with cystic fibrosis, the 508th amino acid of the CFTR protein (a phenylalanine) is missing. A CFTR protein with this change is synthesized correctly, and it can transport ions correctly, but it never reaches the plasma membrane to do its job.

Sergei Bannykh and his coworkers developed a procedure to measure the relative amounts of the CFTR protein localized in different regions of a cell. They compared the pattern of CFTR distribution in normal cells with the pattern in CFTR-mutated cells. A summary of their results is shown in the adjacent figure.

1. Which organelle contains the least amount of CFTR protein in normal cells? In CF cells? Which contains the most?

2. In which organelle is the amount of CFTR protein in CF cells closest to the amount in normal cells?

3. Where is the mutated CFTR protein getting held up?

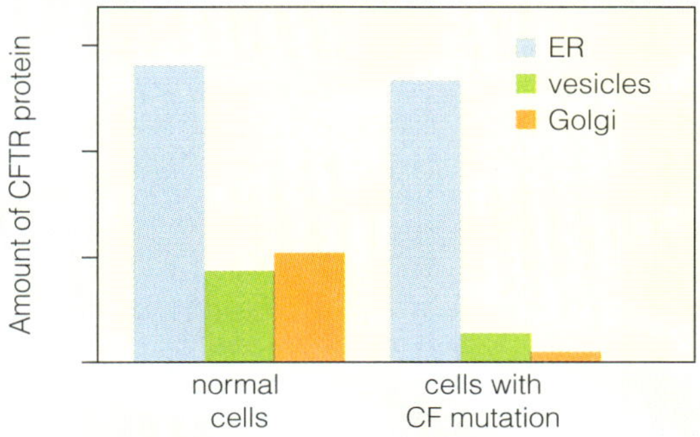

Comparison of the amounts of CFTR protein associated with endoplasmic reticulum (*blue*), vesicles traveling from ER to Golgi (*green*), and Golgi bodies (*orange*). The patterns of CFTR distribution in normal cells, and cells with the most common cystic fibrosis mutation, were compared.

Pittler Study: Effects of Artichoke Extract on Hangovers

Ethanol is a toxin, so it makes sense that drinking it can cause various symptoms of poisoning—headache, stomach ache, nausea, fatigue, impaired memory, dizziness, tremors, and diarrhea, among other ailments. All are symptoms of hangover, the common word for what happens as the body is recovering from a bout of heavy drinking.

The most effective treatment for a hangover is to avoid drinking in the first place. Folk remedies (such as aspirin, coffee, bananas, more alcohol, honey, barley grass, pizza, milkshakes, glutamine, raw eggs, charcoal tablets, or cabbage) abound, but few have been studied scientifically. In 2003, Max Pittler and his colleagues tested one of them. The researchers gave 15 participants an unmarked pill containing either artichoke extract or a placebo (an inactive substance) just before or after drinking enough alcohol to cause a hangover. The results are shown in the chart.

Participant (Age, Gender)	Severity of Hangover	
	Artichoke Extract	Placebo
1 (34, F)	1.9	3.8
2 (48, F)	5.0	0.6
3 (25, F)	7.7	3.2
4 (57, F)	2.4	4.4
5 (34, F)	5.4	1.6
6 (30, F)	1.5	3.9
7 (33, F)	1.4	0.1
8 (37, F)	0.7	3.6
9 (62, M)	4.5	0.9
10 (36, M)	3.7	5.9
11 (54, M)	1.6	0.2
12 (37, M)	2.6	5.6
13 (53, M)	4.1	6.3
14 (48, F)	0.5	0.4
15 (32, F)	1.3	2.5

Results of a study that tested artichoke extract as a hangover preventive. All participants were tested once with the placebo and once with the extract, with a week interval between. Each rated the severity of 20 hangover symptoms on a scale of 0 (not experienced) to 5 ("as bad as can be imagined"). The 20 ratings were averaged as a single, overall rating, which is listed here.

1. How many participants experienced a hangover that was worse with the placebo than with the artichoke extract?

2. How many participants experienced a worse hangover with the artichoke extract?

3. Calculate the numbers you counted in questions 1 and 2 as a percentage of the total number of participants. How much difference is there between the percentages?

4. Does this data support the hypothesis that artichoke extract is an effective hangover treatment? Why or why not?

Tilman Study: Energy Efficiency of Biofuels from Corn, Soy, and Prairie Grasses

Most corn is grown intensively in vast swaths, which means farmers who grow it use fertilizers and pesticides, both of which are often made from fossil fuels. Corn is an annual plant, and yearly harvests tend to cause runoff that depletes soil and pollutes rivers.

In 2006, David Tilman and his colleagues published the results of a 10-year study comparing the net energy output of various biofuels. The researchers grew a mixture of native perennial grasses without irrigation, fertilizer, pesticides, or herbicides, in sandy soil that was so depleted by intensive agriculture that it had been abandoned. They measured the usable energy in biofuels made from the grasses, from corn, and from soy. They also measured the energy it took to grow and produce each kind of biofuel.

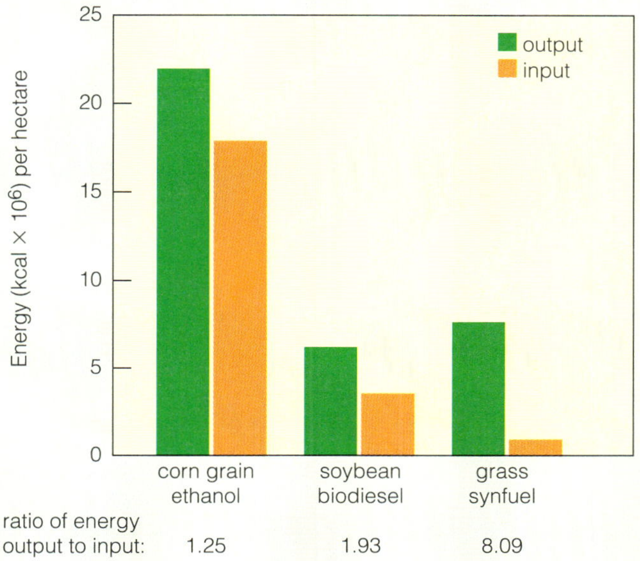

Energy inputs and outputs of biofuels from different sources: corn and soy grown on fertile farmland, and grassland plants grown in infertile soil. One hectare is about 2.5 acres.

1. About how much energy did ethanol produced from one hectare of corn yield? How much energy did it take to grow and produce that ethanol?

2. Which biofuel tested had the highest ratio of energy output to energy input?

3. Which of the three crops would require the least amount of land to produce a given amount of biofuel energy?

Kuruvilla Study: Mitochondrial Abnormalities in TF Patients

Tetralogy of Fallot (TF) is a genetic disorder characterized by four major malformations of the heart. The circulation of blood is abnormal, so TF patients have too little oxygen in their blood. Inadequate oxygen levels result in damaged mitochondrial membranes, which in turn cause cells to self-destruct. In 2004, Sarah Kuruvilla and her colleages looked at abnormalities in the mitochondria of heart muscle in TF patients. Some of their results are shown in the figure to the right.

1. Which abnormality was most strongly associated with TF?

2. Can you make any correlations between blood oxygen content and mitochondrial abnormalities in these patients?

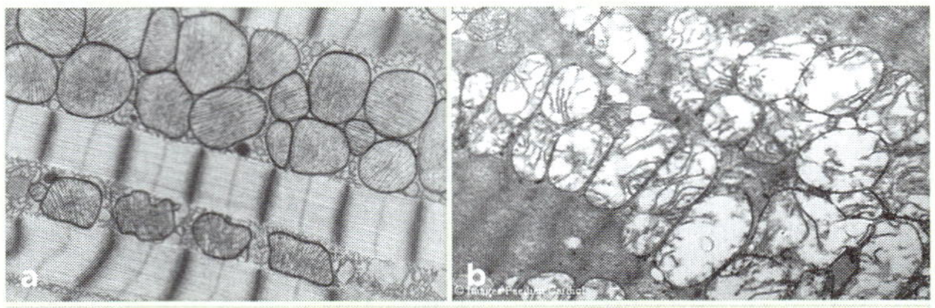

Patient (age)	SPO_2 (%)	Mitochondrial Abnormalities in TF: Number	Shape	Size	Broken
1 (5)	55	+	+	–	–
2 (3)	69	+	+	–	–
3 (22)	72	+	+	–	–
4 (2)	74	+	+	–	–
5 (3)	76	+	+	–	+
6 (2.5)	78	+	+	–	+
7 (1)	79	+	+	–	–
8 (12)	80	+	–	+	–
9 (4)	80	+	+	–	–
10 (8)	83	+	–	+	–
11 (20)	85	+	+	–	–
12 (2.5)	89	+	–	+	–

c

Mitochondrial changes in Tetralogy of Fallot (TF). (**a**) Normal heart muscle. Many mitochondria between the fibers provide muscle cells with ATP for contraction. (**b**) Heart muscle from a person with TF shows swollen, broken mitochondria.

(**c**) Mitochondrial abnormalities in TF patients. SPO_2 is oxygen saturation of the blood. A normal value of SPO_2 is 96%. Abnormalities seen are indicated by "+" signs.

HeLa Cell Chromosome Number

Despite their notorious ability to contaminate other cell lines, HeLa cells continue to be an extremely useful tool in cancer research. One early finding was that HeLa cells vary in chromosome number. The panel of chromosomes in the figure, originally published in 1989, shows all of the chromosomes in a single metaphase HeLa cell.

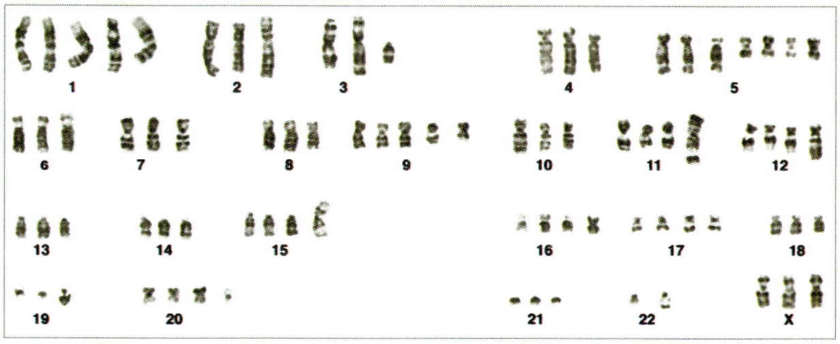

Chromosomes of a HeLa cell line.

1. What is the chromosome number of this HeLa cell?

2. How many extra chromosomes do these cells have, compared to normal human cells?

3. Can you tell these cells came from a female? How?

Case Western University Study: Meiotic Abnormalities Linked to BPA Exposure

In 1998, researchers at Case Western University were studying meiosis in mouse oocytes when they saw an unexpected and dramatic increase of abnormal meiosis events. Improper segregation of chromosomes during meiosis is one of the main causes of human genetic disorders.

The researchers discovered that the spike in meiotic abnormalities started immediately after the mouse facility's plastic cages and water bottles were washed in a new, alkaline detergent. The detergent had damaged the plastic, which began to leach bisphenol A (BPA). BPA is a synthetic chemical that mimics estrogen, a hormone. BPA is used to manufacture polycarbonate plastic items (including baby bottles and water bottles) and epoxies (including the coating on the inside of metal cans of food).

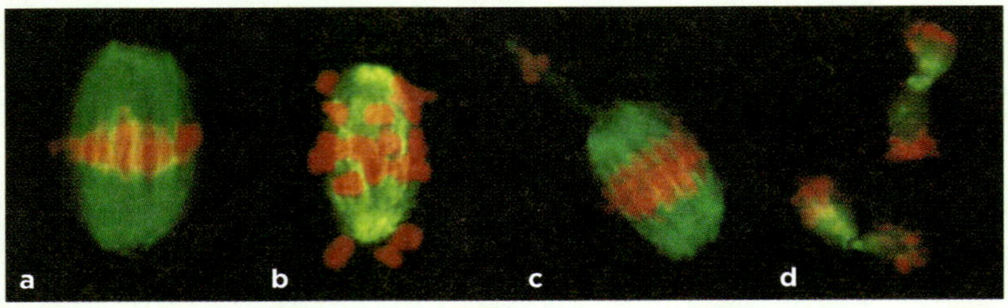

Caging materials	Total number of oocytes	Abnormalities
Control: New cages with glass bottles	271	5 (1.8%)
Damaged cages with glass bottles		
Mild damage	401	35 (8.7%)
Severe damage	149	30 (20.1%)
Damaged bottles	197	53 (26.9%)
Damaged cages with damaged bottles	58	24 (41.4%)

Meiotic abnormalities associated with exposure to damaged plastic caging. Fluorescent micrographs show nuclei of single mouse oocytes in metaphase I. (**a**) Normal metaphase; (**b–d**) examples of abnormal metaphase. Chromosomes are *red;* spindle fibers are *green.*

1. What percentage of mouse oocytes displayed abnormalities of meiosis with no exposure to damaged caging?

2. Which group of mice showed the most meiotic abnormalities in their oocytes?

3. What is abnormal about metaphase I as it is occurring in the oocytes shown in Figure *b*, *c*, and *d*?

Skin Color of Indigenous Peoples Correlated to Regional Incident UV Radiation

A 2000 study measured average skin color of people native to more than fifty regions, and correlated them to the amount of UV radiation received in those regions. Some of their results are shown here.

1. Which country receives the most UV radiation? The least?

2. The people native to which country have the darkest skin? The lightest?

3. According to this data, how does the skin color of indigenous peoples correlate with the amount of UV radiation incident in their native regions?

Country	Skin Reflectance	UVMED
Australia	19.30	335.55
Kenya	32.40	354.21
India	44.60	219.65
Cambodia	54.00	310.28
Japan	55.42	130.87
Afghanistan	55.70	249.98
China	59.17	204.57
Ireland	65.00	52.92
Germany	66.90	69.29
Netherlands	67.37	62.58

Skin color of indigenous peoples and regional incident UV radiation. Skin reflectance measures how much light of 685 nanometers wavelength is reflected from skin; UVMED is the annual average UV radiation received at Earth's surface.

Coryell Study: Genetic Relationship between Mood Disorders and Intelligence

A study in 1989 looked for a genetic relationship between mood disorders and intelligence. William Coryell and his colleagues found people with bipolar disorder, then tallied how many of their immediate family members had graduated from college. Some of the results are shown here.

1. Which group of people, those with bipolar disorder or those without bipolar disorder, had the largest percentage of fathers with a college degree?

2. According to this data, if you had bipolar disorder, which one of your immediate relatives would be more likely to have a college degree? Which would be the least likely?

3. Are relatives of people with or without bipolar disorder more likely to graduate from college?

Proportion of College Graduates Among Relatives of People with Bipolar Disorder

Relative	Nonbipolar (%)	Bipolar (%)
Father	16.1	27.3
Mother	10.5	18.2
Brother	23.9	38.7
Sister	16.1	31.8
Grandfather	7.6	13.0
Grandmother	5.1	7.1

Proportion of immediate relatives of nonbipolar and bipolar people that graduated from college.

Hershey–Chase Experiment

The graph to the right is reproduced from Alfred Hershey and Martha Chase's 1952 publication that showed DNA is the hereditary material of bacteriophage. The data are from two experiments in which bacteriophage DNA and protein were labeled with radioactive tracers and allowed to infect bacteria. The virus–bacteria mixtures were whirled in a blender to dislodge the two, and the tracers were tracked inside and outside of the bacteria.

1. Before blending, what percentage of ^{35}S was outside the bacteria? What percentage was inside? What percentage of ^{32}P was outside the bacteria? What percentage was inside?

2. After 4 minutes in the blender, what percentage of ^{35}S was outside the bacteria? What percentage was inside? What percentage of ^{32}P was outside the bacteria? What percentage was inside?

3. How did the researchers know that the radioisotopes in the fluid came from outside the bacterial cells (extracellular) and not from bacteria that had broken apart?

4. The extracellular concentration of which isotope, ^{35}S or ^{32}P, increased the most with blending? DNA contains much more phosphorus than do proteins; proteins contain much more sulfur than do DNA. Do these results imply that the viruses inject DNA or protein into bacteria? Why?

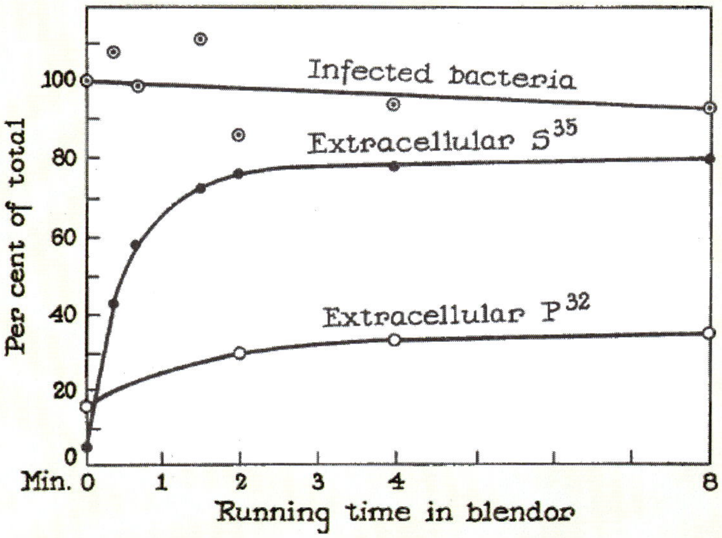

Detail of Alfred Hershey and Martha Chase's publication describing their experiments with bacteriophage. "Infected bacteria" refers to the percentage of bacteria that survived the blender.

From the Journal of General Physiology, *36(1), September 20, 1952: "Independent Functions of Viral Protein and Nucleic Acid in Growth of Bacteriophage."*

Serra Study: Mutations in Individual Neurofibromatosis Tumors

About one out of 3,500 people carry a mutation that affects the product of the NF1 gene, which is a tumor suppressor. People who are heterozygous for one of these mutations have neurofibromatosis, an autosomal dominant genetic disorder. Among other problems, soft, fibrous tumors (neurofibromas) form in the skin and nervous system. The homozygous condition may be lethal.

Most mutations associated with neurofibromatosis result in defective splicing of the gene's 60 exons. Each neurofibroma typically arises from a new mutation that disrupts the individual's one functional allele. In a 1997 study, Eduard Serra and his colleagues tested several tumors from an individual with the disorder for such mutations.

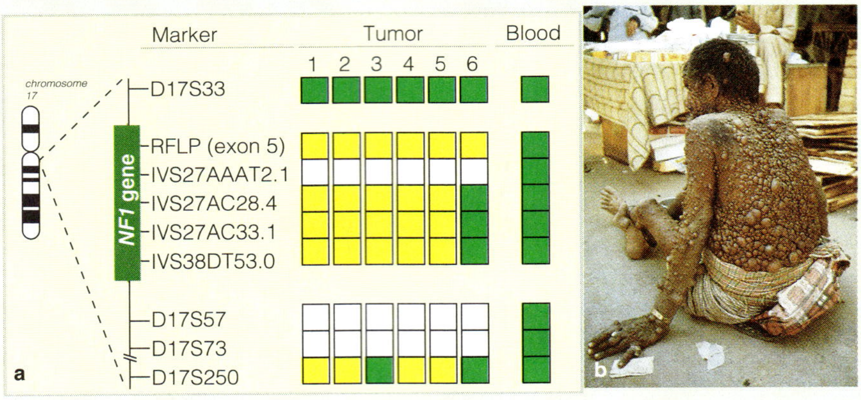

Neurofibromatosis. (**a**) Genetic analysis of six tumors from a single individual affected by neurofibromatosis. Each tumor was checked for the presence of nine nucleotide sequences (markers) in or near the *NF1* gene.

For each tumor (1–6), *green* boxes indicate that the marker is present; *yellow* boxes indicate the marker is missing; *white* boxes indicate inconclusive results. Blood was also tested as a control.

(**b**) An individual affected by neurofibromatosis.

1. Which tumors are missing marker D17S250? Is this sequence inside or outside of the *NF1* gene?

2. In four of these six tumors, the entire large arm of chromosome 17 was deleted. Which four?

3. People affected by neurofibromatosis are 200 to 500 times more likely to develop malignant tumors than unaffected people. Why do you think that is the case?

BRCA Mutations in Women Diagnosed with Breast Cancer

Investigating a correlation between specific cancer-causing mutations and risk of mortality in humans is challenging, in part because each cancer patient is given the best treatment available at the time. There are no "untreated control" cancer patients, and the idea of what treatments are the best changes quickly as new drugs become available and new discoveries are made.

The chart to the right shows a study in which 442 women who had been diagnosed with breast cancer were checked for *BRCA* mutations, and their treatments and progress were followed over several years. All of the women in the study had at least two affected close relatives, so their risk of developing breast cancer due to an inherited factor was estimated to be greater than that of the general population.

BRCA Mutations in Women Diagnosed With Breast Cancer

	BRCA1	*BRCA2*	No *BRCA* Mutation	Total
Total number of patients	89	35	318	442
Avg. age at diagnosis	43.9	46.2	50.4	
Preventive mastectomy	6	3	14	23
Preventive oophorectomy	38	7	22	67
Number of deaths	16	1	21	38
Percent died	18.0	2.8	6.9	8.6

Results from a 2007 study investigating BRCA mutations in women diagnosed with breast cancer. All women in the study had a family history of breast cancer.

Some of the women underwent preventive mastectomy (removal of the noncancerous breast) during their course of treatment. Others had preventive oophorectomy (surgical removal of the ovaries) to prevent the possibility of getting ovarian cancer.

1. According to this study, what is a woman's risk of dying of cancer if two of her close relatives have breast cancer?

2. What is her risk of dying of cancer if she carries a mutated *BRCA1* gene?

3. Is a *BRCA1* or *BRCA2* mutation more dangerous in breast cancer cases?

4. What other data would you have to see in order to make a conclusion about the effectiveness of preventive surgeries?

Tabuchi Study: Enhanced Spatial Learning Ability in Mice with Autism Mutation

Autism is a neurobiological disorder with a range of symptoms that include impaired social interactions, stereotyped patterns of behavior such as hand-flapping or rocking, and, occasionally, greatly enhanced intellectual abilities.

Autism may have a genetic basis. Some autistic people have a mutation in neuroligin 3, a type of cell adhesion protein that connects brain cells. One mutation changes amino acid 451 from arginine to cysteine.

Mouse and human neuroligin 3 are very similar. In 2007, Katsuhiko Tabuchi and his colleagues genetically modified mice to carry the same arginine-to-cysteine substitution in their neuroligin 3. The mutation caused an increase in transmission of some types of signals between brain cells. Mice with the mutation had impaired social behavior, and, unexpectedly, enhanced spatial learning ability.

1. Did the modified or the unmodified mice learn the location of the platform faster in the first test?

2. Which mice learned faster the second time around?

3. Which mice showed the greatest improvement in memory between the first and the second test?

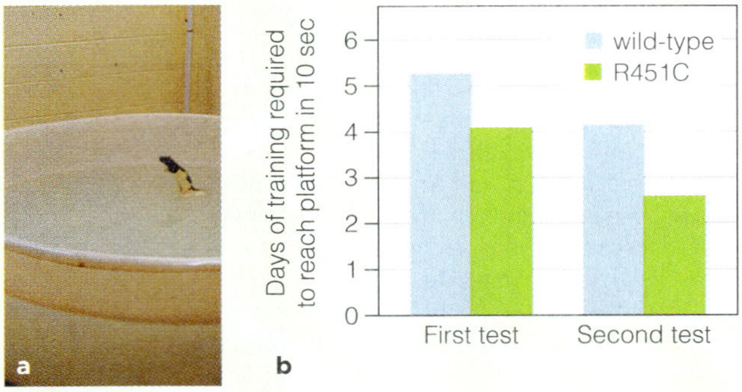

Enhanced spatial learning ability in mice with a mutation in neuroligin 3 (R451C), compared with unmodified (wild-type) mice. The mice were tested in a water maze, in which a platform is submerged a few millimeters below the surface of a deep pool of warm water (**a**). The platform, which is not visible to swimming mice, was moved for the second test.

Mice do not particularly enjoy swimming, so they locate a hidden platform as fast as they can. When tested again, they can remember its location by checking visual cues around the edge of the pool. How quickly they remember the platform's location is a measure of their spatial learning ability (**b**).

Alvarez Study: Abundance of Iridium in the K–T Boundary

In the late 1970s, geologist Walter Alvarez was investigating the composition of the 1-centimeter-thick layer of clay that marks the K–T boundary all over the world. He asked his father, Nobel Prize–winning physicist Luis Alvarez, to help him analyze the elemental composition of the layer. The Alvarezes and their colleagues, chemists Frank Asaro and Helen Michel, tested the layer in Italy and in Denmark. The researchers discovered that the K–T boundary layer had a much higher iridium content than the surrounding rock layers.

Iridium belongs to a group of elements (Appendix IV) that are much more abundant in asteroids and other solar system materials than they are in Earth's crust. The Alvarez group concluded that the K–T boundary layer must have originated with extraterrestrial material. They calculated that an asteroid 14 kilometers (8.7 miles) in diameter would contain enough iridium to account for the iridium in the K–T boundary layer.

Sample Depth	Average Abundance of Iridium (ppb)
+ 2.7 m	< 0.3
+ 1.2 m	< 0.3
+ 0.7 m	0.36
boundary layer	41.6
– 0.5 m	0.25
– 5.4 m	0.30

Abundance of iridium in and near the K–T boundary layer in Stevns Klint, Denmark. Many rock samples taken from above, below, and at the boundary layer were tested for iridium content. Depths are given as meters above or below the boundary layer.

The iridium content of an average Earth rock contains 0.4 parts per billion (ppb) of iridium. An average meteorite contains about 550 parts per billion of iridium. Photo (above *right*): Luis and Walter Alvarez with a section of the iridium layer.

1. What was the iridium content of the K–T boundary layer?

2. How much higher was the iridium content of the boundary layer than the sample taken 0.7 meters above the layer?

Kohn Study: Resistance to Rodenticides in Wild Rat Populations

Beginning in 1990, rat infestations in northwestern Germany started to intensify despite continuing use of rodenticides. In 2000, Michael H. Kohn and his colleagues analyzed the genetics of wild rat populations around Munich. For part of their research, they trapped wild rats in five towns, and tested those rats for resistance to warfarin and the second-generation anticoagulant bromadiolone. The results are shown in the figure to the right.

1. In which of the five towns were most of the rats susceptible to warfarin?

2. Which town had the highest percentage of anticoagulant-resistant wild rats?

3. What percentage of rats in Olfen were resistant to warfarin?

4. In which town do you think the application of bromadiolone was most intensive?

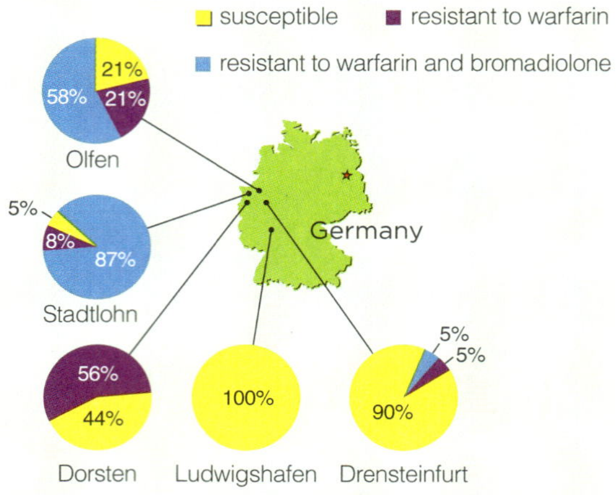

Resistance to anticoagulant rodenticides in wild populations of rats in Germany, 2000.

Fleischer Study: Honeycreeper Phylogeny

The poouli (*Melamprosops phaeosoma*) was discovered in 1973 by a group of students from the University of Hawaii. Its membership in the honeycreeper clade was—and continues to be—controversial, mainly because its appearance, odor, and behavior are so different from other honeycreepers. It particularly lacks the "old tent" scent of other honeycreepers.

A study published in 2001 by Robert Fleischer and his colleagues tried to clarify the relationships of *Melamprosops* to other honeycreepers by compared bone morphology and DNA sequences. Some of their data, and the resulting cladogram, are shown here.

1. According to the cladogram, which species is/are most closely related to the poouli? The one(s) most closely related to the ancestor of the honeycreepers?

2. Is any honeycreeper species more closely related to the ancestral species than the poouli?

3. Not counting unresolved bases, how many differences are between the poouli and *Melospiza georgiana* sequences? The poouli and *Himatione sanguinea* sequences?

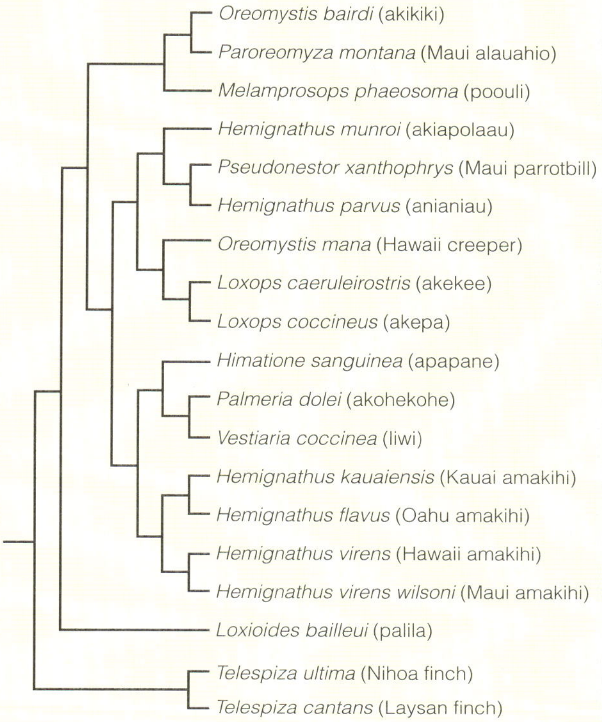

DNA sequence comparison of part of the cytochrome *b* control sequence from different honeycreeper species (*below*), and cladogram of honeycreeper phylogeny that resulted (*right*). The top sequence is from an outgroup species; N indicates an unresolved nucleotide. Differences from the *Melamprosops* sequence are indicated in *red*; gaps in the alignment are dashed.

Melospiza georgiana (swamp sparrow)	TAGCCACGACACCTTATTATGAA-CCACTAGTGA-A-AACACTCCCGTAGGTATATTCAATAGATAG
Paroroeomyza montana	TCACCAAGACGATCTATTACGCTACACCAGGGAGGATGGCACTCCCACTGGTATATCCACTTGACAG
Loxioides bailleui	TCACTAAGACAGCTTACTACGCCAACTCACGCGAGAGAGCACTCCCGGTGGTATACTCACTTGATAG
Telespiza cantans	TCACCAAGACGGTTCACTATACCACACCAAGTGAGAGAGCACTCNNNNNGGTATACTCACTTTACAG
Hemignathus parvus	TCACCAAGACGACTTATTGTATCAA-CCAAGAGAGANNGCACTCCCAGTGGTAGATTCTCTTGACAG
Hemignathus kauaiensis	TCACTAAGACGACTTATTATACCCAACCAAGAGAAAAAGCACCCCCCACGGTAAGTCCTCTTGACCG
Himatione sanguinea	TCGCTTAGACGCCTTATTACGCTAAACCATCACAGANNGCACTACTGTTGGTCGATCCTCTTGACAG
Melamprosops phaeosoma	TCACTAAGACACCTTCTTATGTTCCATCAAGAGAGANNGCACNNNNNNNGGTATAGCTTCTTGACAG

Changes in Conditions on Early Earth

Studies of ancient rocks and fossils can reveal changes that have taken place during Earth's existence. The adjacent figure shows how asteroid impacts and composition of the atmosphere are thought to have changed over time. Use this figure and information in the chapter to answer the following questions.

1. Which occurred first, a decline in asteroid impacts, or a rise in the atmospheric level of oxygen?

2. How do modern levels of carbon dioxide and oxygen compare to those at the time when the first cells arose?

3. Which is now more abundant, oxygen or carbon dioxide?

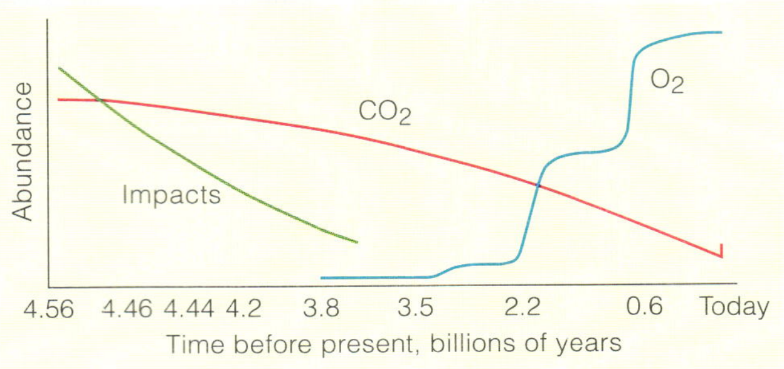

How asteroid impacts (*green*), atmospheric carbon dioxide concentration (*pink*), and oxygen concentration (*blue*) changed over geologic time.

HIV Transmission from Mother to Child

One of the sadder aspects of the AIDS pandemic is transmission of HIV from mother to child. Untreated mothers have a 15 to 30 percent chance of passing on the infection during pregnancy. The virus is also transmitted by breast-feeding.

Since the AIDS pandemic began, there have been more than 8,000 cases of mother-to-child HIV transmission in the United States. In 1993, American physicians began giving antiretroviral drugs to HIV-positive women during pregnancy and treating both mother and infant in the months after birth. Only about 10 percent of mothers were treated in 1993, but by 1999 more than 80 percent got the drugs.

The figure shows the number of AIDS diagnoses among children in the United States. Use the information in this graph to answer the following questions.

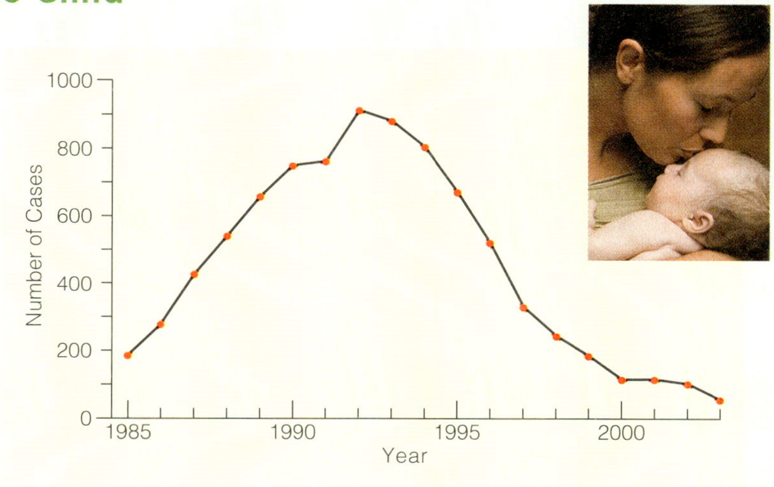

Number of new AIDS diagnoses in the United States per year among children who were exposed to HIV during pregnancy, birth, or by breast-feeding.

1. How did the number of children diagnosed with AIDS change during the late 1980s?

2. What year did new AIDS diagnoses in children peak, and how many children were diagnosed that year?

3. How did the number of AIDS diagnoses change as the use of antiretrovirals in mothers and infants increased?

Jacob Koella Study: Attractiveness of *Plasmodium*-Infected Hosts to Mosquitoes

Parasites sometimes alter their host's behavior in a way that increases their chances of transmission to another host. For example, *Toxoplasma gondii*, the cause of toxoplasmosis, infects rats and makes them less wary of cats.

Dr. Jacob Koella and his associates hypothesized that *Plasmodium* might benefit by making its human host more attractive to hungry mosquitoes when gametocytes were available in the host's blood. Gametocytes are the stage that can be taken up by the mosquito and mature into gametes inside its gut.

To test their hypothesis, the researchers recorded the response of mosquitoes to the odor of *Plasmodium*-infected children and uninfected children over the course of 12 trials on 12 separate days. The figure to the right shows their results.

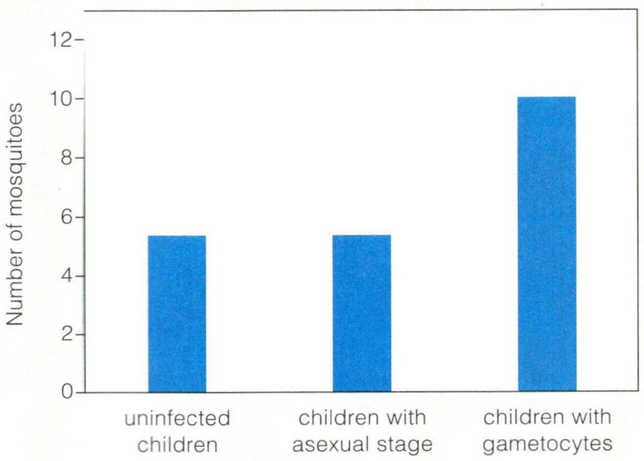

Number of mosquitoes (out of 100) attracted to uninfected children, children harboring the asexual stage of *Plasmodium*, and children with gametocytes in their blood. The bars show the average number of mosquitoes attracted to that category of child over the course of 12 separate trials.

1. On average, which group of children was most attractive to mosquitoes?

2. Which group of children averaged the fewest mosquitoes attracted?

3. What percentage of the total number of mosquitoes were attracted to the most attractive group?

4. Did the data support Dr. Koella's hypothesis?

UNFAO Report: On Deforestation

The U.N. Food and Agriculture Organization (UNFAO) recognizes the importance of forests to human populations and keeps track of forest abundance. The figure shows UNFAO data on the amount of forest throughout the world in 1990, 2000, and 2005.

Region	Forested Area (in millions of hectares) 1990	2000	2005
Africa	699	656	635
Asia	574	567	572
Central America	28	24	22
Europe	989	988	1001
North America	678	678	677
Oceania	233	208	206
South America	891	853	832
World total	4,077	3,988	3,952

Changes in forested area by region from 1990 to 2005. One hectare is 2.47 acres. The full report on the world's forests is online at www.fao.org/forestry/en/.

1. How many hectares of forested land were there in North America in 2005?

2. In which region(s) did the amount of forested land increase between 1990 and 2005?

3. How many hectares of forest did the world lose between 1990 and 2005?

4. In 2002, China embarked on an ambitious campaign to add 76 million hectares of trees over a ten-year period. Do you see any indication that this campaign is successful?

Effect of Logging Practices on Tree Loss to a Fungal Pathogen

The club fungus *Armillaria ostoyae* infects living trees and acts as a parasite, withdrawing nutrients from them. If the tree dies, the fungus continues to dine on its remains. Fungal hyphae grow out from the roots of infected trees and roots of dead stumps. If these hyphae contact roots of a healthy tree, they can invade and cause a new infection.

Canadian forest pathologists hypothesized that remov-ing stumps after logging could help prevent tree deaths. To test this hypothesis, they carried out an experiment. In half of a forest they removed stumps after logging. In a control area, they left stumps behind. For more than 20 years, they recorded tree deaths and whether *A. ostoyae* caused them. The figure to the right shows the results.

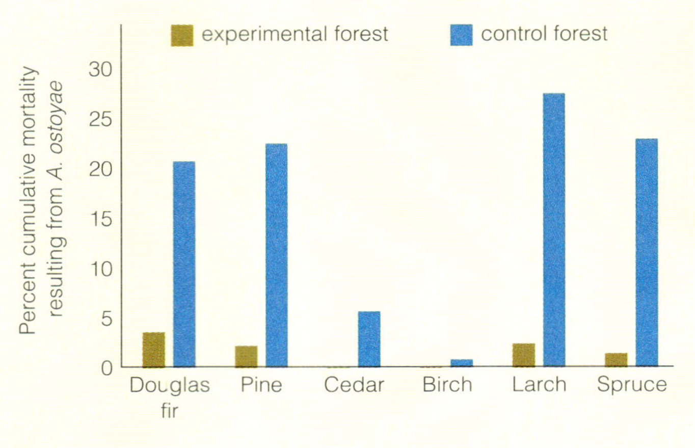

Results of a long-term study of how logging practices affect tree deaths caused by the fungus *A. ostoyae*. In the experimental forest, whole trees—including stumps—were removed (*brown* bars). The control half of the forest was logged conventionally, with stumps left behind (*blue* bars).

1. Which tree species was most often killed by *A. ostoyae* in control forests? Which was least affected by the fungus?

2. For the most affected species, what percentage of deaths did *A. ostoyae* cause in control and in experimental forests?

3. Looking at the overall results, do the data support the hypothesis? Does stump removal reduce effects of *A. ostoyae*?

Survival of Horseshoe Crabs after Bleeding

Atlantic horseshoe crabs, *Limulus polyphemus*, have long been ecologically important. For more than a million years, their eggs have fed migratory shorebirds. More recently, people began to harvest horseshoe crabs for use as bait. More recently still, people started using horseshoe crab blood to test injectable drugs for potentially deadly bacterial toxins. To keep horseshoe crab populations stable, blood is extracted from captured animals, which are then returned to the wild. Concerns about the survival of animals after bleeding led researchers to do an experiment. They compared survival of animals captured and maintained in a tank with that of animals captured, bled, and kept in a similar tank. The figure to the right shows the results.

	Control Animals		Bled Animals	
Trial	Number of crabs	Number that died	Number of crabs	Number that died
1	10	0	10	0
2	10	0	10	3
3	30	0	30	0
4	30	0	30	0
5	30	1	30	6
6	30	0	30	0
7	30	0	30	2
8	30	0	30	5
Total	200	1	200	16

Mortality of young male horseshoe crabs kept in tanks during the 2 weeks after their capture. Blood was taken from half the animals on the day of their capture. Control animals were handled, but not bled. This procedure was repeated 8 times with different sets of horseshoe crabs.

1. In which trial did the most control crabs die? In which did the most bled crabs die?

2. Looking at the overall results, how did the mortality of the two groups differ?

3. Based on these results, would you conclude that bleeding harms horseshoe crabs more than capture alone does?

Trends in Primate Evolution

One trend in primate evolution involved changes in life history traits, such as length of infancy and the time it takes to reach adulthood.

The figure compares five primate lineages, from most ancient to most recent. It graphs life spans and years spent as "infants" when ongoing maternal care is required. It shows time spent as subadults, when individuals are no longer dependent on their mother for care, but have not yet begun to breed. It shows the length of the reproductive years, and the length of time lived after reproductive years have passed.

1. What is the average life span for a lemur? A gibbon?
2. Which group reaches adulthood most quickly?
3. Which group has the longest expanse of reproductive years?
4. Which groups survive past their reproductive years?

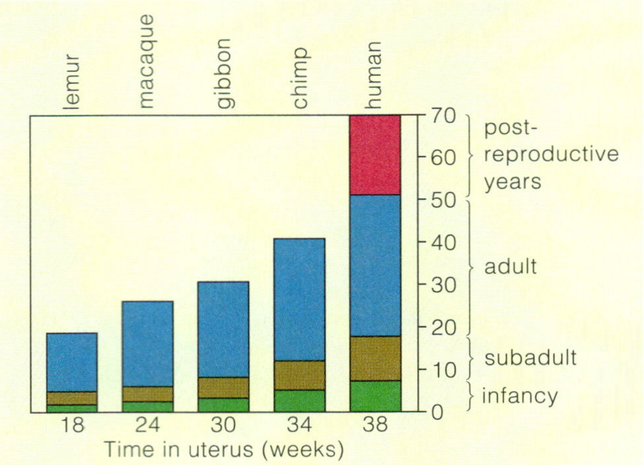

Trend toward longer life spans and greater dependency of offspring on adults for five primate lineages.

Heat Index (Apparent Temperature)

As part of ongoing efforts to prevent heat-related illness, the National Weather Service has devised a heat index (HI) to alert people to the risks of high temperature coupled with high humidity. The heat index is sometimes referred to as the "apparent temperature." It tells you what the temperature feels like, given the level of relative humidity. The higher the HI value, the higher the heat disorder risk with prolonged exposure or with exertion.

The figure shows the heat index chart. The maximum possible value is 137. *Gold* indicates temperatures near the danger level, *orange* indicates danger, and *pink* means extreme danger.

Temp (°F)	Relative humidity (%)												
	40	45	50	55	60	65	70	75	80	85	90	95	100
110	136												
108	130	137											
106	124	130	137										
104	119	124	131	137									
102	114	119	124	130	137								
100	109	114	118	124	129	136							
98	105	109	113	117	123	128	134						
96	101	104	108	112	116	121	126	132					
94	97	100	103	106	110	114	119	124	129	135			
92	94	96	99	101	105	108	112	116	121	126	131		
90	91	93	95	97	100	103	106	109	113	117	122	127	132
88	88	89	91	93	95	98	100	103	106	110	113	117	121
86	85	87	88	89	91	93	95	97	100	102	105	108	112
84	83	84	85	86	88	89	90	92	94	96	98	100	103
82	81	82	83	84	84	85	86	88	89	90	91	93	95
80	80	80	81	81	82	82	83	84	84	85	86	86	87

Heat index (HI) chart.

1. What is the heat index on a day when the temperature is 96°F and the relative humidity is 45 percent?

2. What is the heat index on a day when the temperature is 96°F and the relative humidity is 75 percent?

3. How does the danger level indicated by these two heat index values compare?

4. What is the lowest temperature that, when coupled with 100% relative humidity, can cause extreme danger?

Grissino-Mayer Study: Dendroclimatology Data from El Malpais National Monument

Douglas fir trees (*Pseudotsuga menziesii*) are exceptionally long-lived, and particularly responsive to rainfall levels. Researcher Henri Grissino-Mayer sampled Douglas firs in El Malpais National Monument, in west central New Mexico. Pockets of vegetation in this site have been surrounded by lava fields for about 3,000 years, so they have escaped wildfires, grazing animals, agricultural activity, and logging. Grissino-Mayer compiled tree ring data from old, living trees, and dead trees and logs to generate a 2,129-year annual precipitation record.

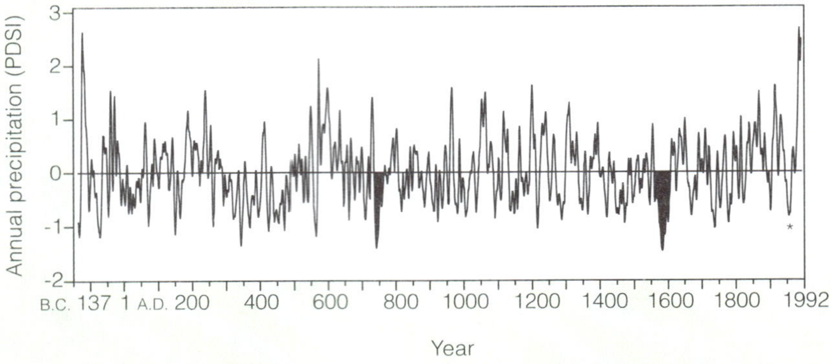

A 2,129-year annual precipitation record complied from tree rings in El Malpais National Monument, New Mexico. Data was averaged over 10-year intervals; graph correlates with other indicators of rainfall collected in all parts of North America. PDSI: Palmer Drought Severity Index: 0, normal rainfall; increasing numbers mean increasing excess of rainfall; decreasing numbers mean increasing severity of drought.

* A severe drought contributed to a series of catastrophic dust storms that turned the midwestern United States into a "dust bowl" between 1933 and 1939.

1. The Mayan civilization began to suffer a massive loss of population around 770 A.D. Do these tree ring data reflect a drought condition at this time? If so, was that condition relatively more or less severe than the "dust bowl" drought?

2. One of the worst population catastrophes ever recorded occurred in Mesoamerica between 1519 and 1600 A.D., when approximately 22 million people native to the region died. According to these data, which period between 137 B.C. and 1992 had the most severe drought? How long did that particular drought last?

Doty Study: Usefulness of Cytochrome P450 Transgenic Plants for Phytoremediation of TCE

Plants used for phytoremediation take up organic pollutants from the soil or air, then transport the chemicals to plant tissues, where they are stored or broken down. Researchers are now designing transgenic plants with enhanced ability to take up or break down toxins.

In 2007, Sharon Doty and her colleagues published the results of their efforts to design plants useful for phytoremediation of soil and air containing organic solvents. The researchers used *Agrobacterium tumefaciens* to deliver a mammalian gene into poplar plants. The gene encodes cytochrome P450, a type of heme-containing enzyme involved in the metabolism of a range of organic molecules, including solvents such as TCE. The results of one of the researchers' tests on these transgenic plants are shown in the figure to the right.

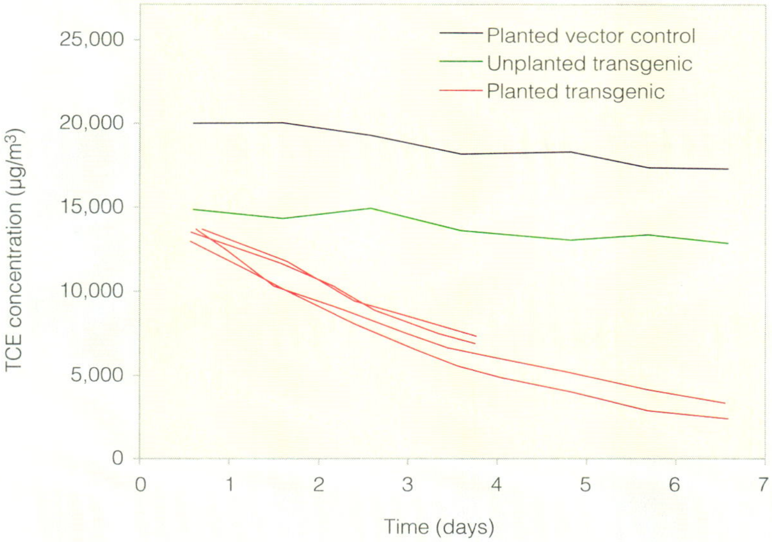

Results of tests on transgenic poplar trees. Planted trees were incubated in sealed containers with an initial 15,000 micrograms of TCE (trichloroethylene) per cubic meter of air. Samples of the air in the containers were taken daily and measured for TCE content. Controls included a tree transgenic for a *Ti* plasmid with no cytochrome P450 in it (vector control), and a bare-root transgenic tree (one that was not planted in soil).

1. How many transgenic plants did the researchers test?

2. In which group did the researchers see the slowest rate of TCE uptake? The fastest?

3. On day 6, what was the difference between the TCE content of air around transgenic plants and that around vector control plants?

4. Assuming no other experiments were done, what two explanations are there for the results of this experiment? What other control might the researchers have used?

Search for Pollinators of *M. depressa* Plants

Massonia depressa is a low-growing succulent plant native to the desert of South Africa. The dull-colored flowers of this monocot develop at ground level, have tiny petals, emit a yeasty aroma, and produce a thick, jelly-like nectar. These features led researchers to suspect that desert rodents such as gerbils pollinate this plant.

To test their hypothesis, the researchers trapped rodents in areas where *M. depressa* grows and checked them for pollen. They also put some plants in wire cages that excluded mammals, but not insects, to see whether fruits and seeds would form in the absence of rodents. The results are shown in the charts below.

1. How many of the 13 captured rodents showed some evidence of pollen from *M. depressa*?

2. Would this evidence alone be sufficient to conclude that rodents are the main pollinators for this plant?

3. How did the average number of seeds produced by caged plants compare with that of control plants?

4. Do these data support the hypothesis that rodents are required for pollination of *M. depressa*? Why or why not?

The dull, petal-less, ground-level flowers of *Massonia depressa* are accessible to rodents, who push their heads through the stamens to reach the nectar. Note the pollen on the gerbil's snout.

a

Type of rodent	Number caught	# with pollen on snout	# with pollen in feces
Namaqua rock rat	4	3	2
Cape spiny mouse	3	2	2
Hairy-footed gerbil	4	2	4
Cape short-eared gerbil	1	0	1
African pygmy mouse	1	0	0

b

	Mammals allowed access to plants	Mammals excluded from plants
Percentage of plants that set fruit	30.4	4.3
Average number of fruits per plant	1.39	0.47
Average number of seeds per plant	20.0	1.95

Right, results of experiments testing rodent pollination of *M. depressa*. (**a**) Evidence of visits to *M. depressa* by rodents. (**b**) Fruit and seed production of *M. depressa* with and without visits by mammals. Mammals were excluded from plants by wire cages with openings large enough for insects to pass through. 23 plants were tested in each group.

Delphia Study: Predator-Specific Volatile Compounds Produced by Tobacco Plants

In 2007, researchers Casey Delphia, Mark Mescher, and Consuelo De Moraes published a study on the production of different volatile chemicals by tobacco plants (*Nicotiana tabacum*) in response to predation by two types of insects: western flower thrips (*Frankliniella occidentalis*) and tobacco budworms (*Heliothis virescens*). Their results are shown to the right.

1. Which treatment elicited the greatest production of volatiles?
2. Which volatile chemical was produced in the greatest amount? What was the stimulus?
3. Which one of the chemicals tested is most likely produced by tobacco plants in a nonspecific response to predation?
4. Are there any chemicals produced in response to predation by budworms, but not in response to predation by thrips?

Volatile Compound Produced	Treatment					
	C	T	W	WT	HV	HVT
Myrcene	0	0	0	0	17	22
β-Ocimene	0	433	15	121	4,299	5,315
Linalool	0	0	0	0	125	178
Indole	0	0	0	0	74	142
Nicotine	0	0	233	160	390	538
β-Elemene	0	0	0	0	90	102
β-Caryophyllene	0	100	40	124	3,704	6,166
α-Humulene	0	0	0	0	123	209
Sesquiterpene	0	7	0	0	219	268
α-Farnesene	0	15	0	0	293	457
Caryophyllene oxide	0	0	0	0	89	166
Total	0	555	288	405	9,423	13,563

Volatile compounds produced by tobacco plants (*Nicotiana tabacum*) in response to predation by different insects. Groups of plants were untreated (C), attacked by thrips (T), mechanically wounded (W), mechanically wounded and attacked by thrips (WT), attacked by budworms (HV), or attacked by budworms and thrips (HVT). Values are indicated in nanograms/day.

Effect of Cultured Skin Treatment on Diabetic Foot Ulcers

Diabetes is a disorder in which the blood sugar level is not properly controlled. Among other effects, this disorder reduces blood flow to the lower legs and feet. As a result, about 3 million diabetes patients have ulcers, or open wounds that do not heal, on their feet. Each year, about 80,000 require amputations.

Several companies provide cultured cell products designed to promote the healing of diabetic foot ulcers. The figure to the right shows the results of a clinical experiment that tested the effect of a cultured skin product versus standard treatment for diabetic foot wounds. Patients were randomly assigned to either the experimental treatment group or the control group and their progress was monitored for 12 weeks.

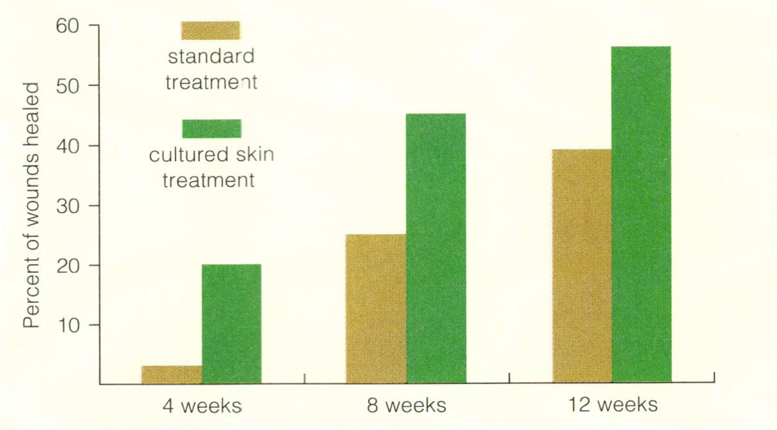

Results of a multicenter study of the effects of standard treatment versus use of a cultured cell product for diabetic foot ulcers. Bars show the percentage of foot ulcers that had completely healed.

1. What percentage of wounds had healed at 8 weeks when treated the standard way? When treated with cultured skin?

2. What percentage of wounds had healed at 12 weeks when treated the standard way? When treated with cultured skin?

3. How early was the healing difference between the control and treatment groups obvious?

Jack Lipton Study: Behavioral Effect of Prenatal Exposure to MDMA

Animal studies are often used to assess effects of prenatal exposure to illicit drugs. For example, Jack Lipton used rats to study the behavioral effect of prenatal exposure to MDMA, the active ingredient in Ecstasy. He injected female rats with either MDMA or saline solution when they were 14 to 20 days pregnant. This is the period when their offsprings' brains were forming. When those offspring were 21 days old, Lipton tested their ability to adjust to a new environment. He placed each young rat in a new cage and used a photobeam system to record how much each rat moved around before settling down. The chart shows his results.

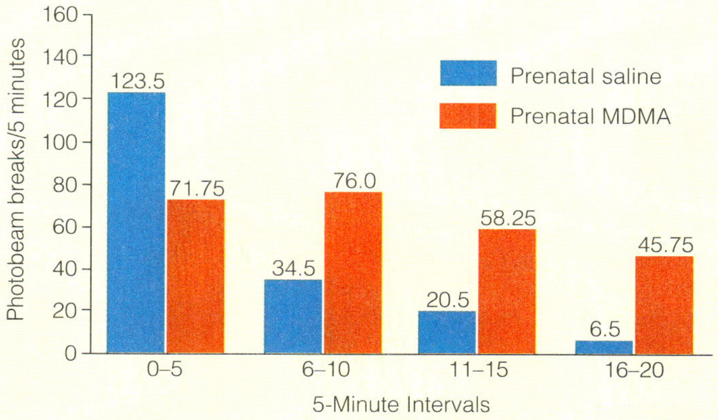

Effect of prenatal exposure to MDMA on activity levels of 21-day-old rats placed in a new cage. Movements were detected when the rat interrupted a photobeam. Rats were monitored at 5-minute intervals for a total of 20 minutes. *Blue* bars are results for rats whose mothers received saline, *red* bars are rats whose mothers received MDMA.

1. Which rats moved around most (caused the most photobeam breaks) during the first 5 minutes in a new cage, those prenatally exposed to MDMA or the controls?

2. How many photobeam breaks did the MDMA-exposed rats make during their second 5 minutes in the new cage?

3. Which rats moved around the most during the last 5 minutes of the study?

4. Does this study support the hypothesis that MDMA affects a developing rat's brain?

Effects of Age and Occupational Noise Exposure

Frequent exposure to noise of a particular pitch can cause loss of hair cells in the part of the cochlea's coil that responds to that pitch. Many workers are at risk for such frequency-specific hearing loss because they work with or around noisy machinery. Taking precautions such as using ear plugs to reduce sound exposure is important. Noise-induced hearing loss can be prevented, but once it occurs it is irreversible. Dead or damaged hair cells are not replaced.

The graph shows the threshold decibel levels at which sounds of different frequencies can be detected by an average 25-year-old carpenter, a 50-year-old carpenter, and a 50-year-old who has not been exposed to on-the-job noise. Sound frequencies are given in hertz (cycles per second). The more cycles per second, the higher the pitch.

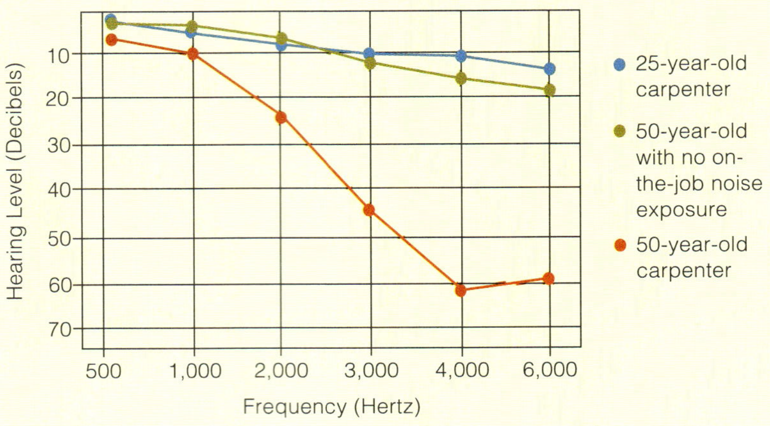

Effects of age and occupational noise exposure. The graph shows the threshold hearing capacities (in decibels) for sounds of different frequencies (given in hertz) in a 25-year-old carpenter (*blue*), a 50-year-old carpenter (*red*), and a 50-year-old who did not have any on-the-job noise exposure (*brown*).

1. Which sound frequency was most easily detected by all three people?

2. How loud did a 1,000-hertz sound have to be for the 50-year-old carpenter to detect it?

3. Which of the three people had the best hearing in the range of 4,000 to 6,000 hertz? Which had the worst?

4. Based on this data, would you conclude that the hearing decline in the 50-year-old carpenter was caused by age or by job-related noise exposure?

Shanna Swann Study: Effect of Agricultural Water Contamination on Humans

Contamination of water by agricultural chemicals affects the reproductive function of some animals. Are there effects on humans? Epidemiologist Shanna Swann and her colleagues studied sperm collected from men in four cities in the United States. The men were partners of women who had become pregnant and were visiting a prenatal clinic, so all were fertile. Of the four cities, Columbia, Missouri, is located in the county with the most farmlands. New York City in New York is in an area with no agriculture.

1. In which cities did researchers record the highest and lowest sperm counts?

2. In which cities did samples show the highest and lowest sperm motility (ability to move)?

3. Aging, smoking, and sexually transmitted diseases adversely affect sperm. Could differences in any of these variables explain the regional differences in sperm count?

4. Do these data support the hypothesis that living near farmlands can adversely affect male reproductive function?

	Location of clinic			
	Columbia, Missouri	Los Angeles, California	Minneapolis, Minnesota	New York, New York
Average age	30.7	29.8	32.2	36.1
Percent nonsmokers	79.5	70.5	85.8	81.6
Percent with history of STD	11.4	12.9	13.6	15.8
Sperm count (million/ml)	58.7	80.8	98.6	102.9
Percent motile sperm	48.2	54.5	52.1	56.4

Data from a study of sperm collected from men who were partners of pregnant women that visited prenatal health clinics in one of four cities. STD stands for sexually transmitted disease.

Clinical Trial of Treatment for Osteogenesis Imperfecta

Tiffany, shown to the right, was born with multiple fractures in her arms and legs. By age six, she had undergone surgery to correct more than 200 bone fractures. Her fragile, easily broken bones are symptoms of osteogenesis imperfecta (OI), a genetic disorder caused by a mutation in a gene for collagen. As bones develop, collagen forms a scaffold for deposition of mineralized bone tissue. The scaffold forms improperly in children with OI. The charts also show the results of an experimental test of a new drug. Treated children, all less than two years old, were compared to similarly affected children of the same age who were not treated with the drug.

Treated child	Vertebral area in cm^2 (Initial)	(Final)	Fractures per year
1	14.7	16.7	1
2	15.5	16.9	1
3	6.7	16.5	6
4	7.3	11.8	0
5	13.6	14.6	6
6	9.3	15.6	1
7	15.3	15.9	0
8	9.9	13.0	4
9	10.5	13.4	4
Mean	11.4	14.9	2.6

Control child	Vertebral area in cm^2 (Initial)	(Final)	Fractures per year
1	18.2	13.7	4
2	16.5	12.9	7
3	16.4	11.3	8
4	13.5	7.7	5
5	16.2	16.1	8
6	18.9	17.0	6
Mean	16.6	13.1	6.3

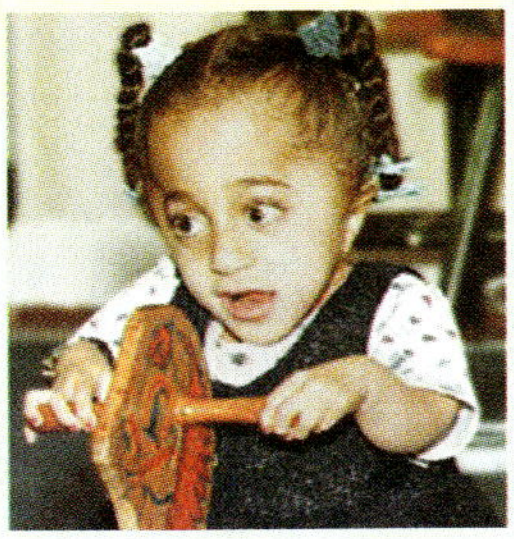

Results of a a clinical trial of a drug treatment for osteogenesis imperfecta (OI), which affects the child shown at *right*. Nine children with OI received the drug. Six others were untreated controls. Surface area of certain vertebrae was measured before and after treatment. Fractures occurring during the 12 months of the trial were also recorded.

1. An increase in vertebral area during the 12-month period of the study indicates bone growth. How many of the treated children showed such an increase?

2. How many of the untreated children showed an increase in vertebral area?

3. How did the rate of fractures in the two groups compare?

4. Do the results shown support the hypothesis that giving young children who have OI this drug, which slows bone breakdown, can increase bone growth and reduce fractures?

Distribution of Stroke Deaths across the United States

Risk of death by stroke is not distributed evenly across the United States. Epidemiologists refer to a swath of states in the Southeast as the "stroke belt" because of the increased incidence of stroke deaths there. By one hypothesis, the high rate of deaths from stroke in this region results largely from a relative lack of access to immediate medical care. Compared to other parts of the country, more stroke-belt residents live in rural settings with few medical services.

The figure to the right compares the rate of stroke deaths in stroke-belt states (Alabama, Arkansas, Georgia, Mississippi, North Carolina, South Carolina, and Tennessee) with that of New York State. It also breaks down the death risk in each region by ethnic group and by sex.

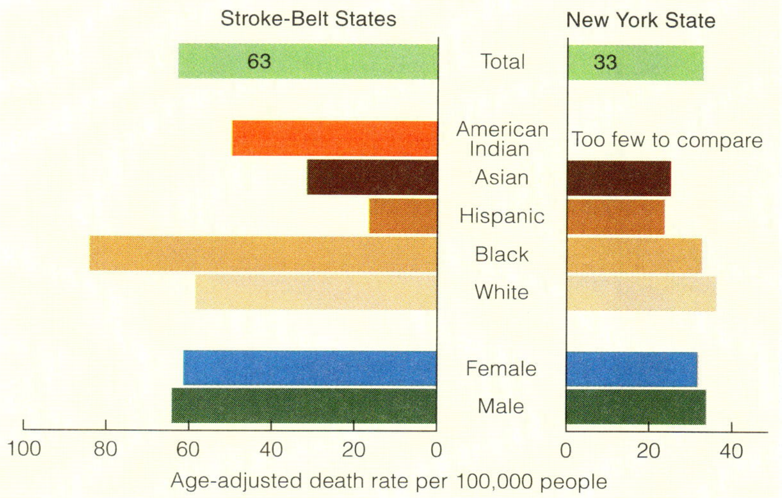

Comparison of the age-adjusted rate of deaths by stroke in Southeastern "stroke-belt" states and in New York State. Source: National Vital Statistics System—Mortality (NVSS-M), NCHS, CDC.

1. How does the rate of stroke deaths among blacks living in the stroke-belt compare with whites in the same region?

2. How does the rate of stroke deaths among blacks living in New York compare with whites in the same region?

3. Which group has the higher rate of stroke deaths, blacks living in New York, or whites living in the stroke belt?

4. Do these data support the hypothesis that poor access to care causes the high rate of death by stroke in the stroke belt?

Khan Study: Cervical Cancer Incidence in HPV-Positive Women

In 2003, Michelle Khan and her coworkers published their findings on a 10-year study in which they followed cervical cancer incidence and HPV status in 20,514 women. All women who participated in the study were free of cervical cancer when the test began. Pap tests were taken at regular intervals, and the researchers used a DNA probe hybridization test to detect the presence of specific types of HPV in the women's cervical cells.

The results are shown here as a graph of the incidence rate of cervical cancer by HPV type. Women who are HPV positive are often infected by more than one type, so the data were sorted into groups based on the women's HPV status ranked by type: either positive for HPV16; or negative for HPV16 and positive for HPV18; or negative for HPV16 and 18 and positive for any other cancer-causing HPV; or negative for all cancer-causing HPV.

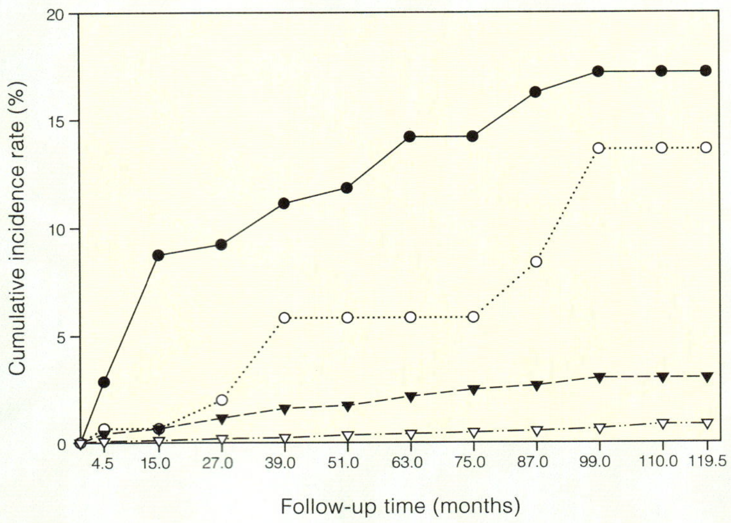

Cumulative incidence rate of cervical cancer correlated with HPV status in 20,514 women aged 16 years and older.

The data were grouped as follows: HPV16 positive (closed circles), or else HPV18 positive (open circles), or else all other cancer-causing HPV types combined (closed triangles). Open triangles: no cancer-causing HPV type was detected.

1. At 110 months into the study, what percentage of women who were not infected with any type of cancer-causing HPV had cervical cancer? What percentage of women who were infected with HPV16 also had cervical cancer?

2. In which group would women infected with both HPV16 and HPV18 fall?

3. Is it possible to estimate from this graph the overall risk of cervical cancer that is associated with infection of cancer-causing HPV of any type?

4. Do these data support the conclusion that being infected with HPV16 or HPV18 raises the risk of cervical cancer?

Risk of Cancer from Radon Exposure

Radon is a colorless, odorless gas emitted by many rocks and soils. It is formed by the radioactive decay of uranium and is itself radioactive. There is some radon in the air almost everywhere, but routinely inhaling a lot of it raises the risk of lung cancer. Radon also seems to increase cancer risk far more in smokers than in nonsmokers. The adjacent chart is an estimate of how radon in homes affects risk of lung cancer mortality. Note that this data shows only the death risk for radon-induced cancers. Smokers are also at risk from lung cancers that are caused by tobacco.

1. If 1,000 smokers were exposed to a radon level of 1.3 pCi/L over a lifetime (the average indoor radon level) how many would die of a radon-induced lung cancer?

2. How high would the radon level have to be to cause approximately the same number of cancers among 1,000 nonsmokers?

3. The risk of dying in a car crash is about 7 out of 1,000. Is a smoker in a home with an average radon level (1.3 pCi/L), more likely to die in a car crash or of radon-induced cancer?

Radon Level (pCi/L)	Risk of Cancer Death From Lifetime Radon Exposure	
	Never Smoked	Current Smokers
20	36 out of 1,000	260 out of 1,000
10	18 out of 1,000	150 out of 1,000
8	15 out of 1,000	120 out of 1,000
4	7 out of 1,000	62 out of 1,000
2	4 out of 1,000	32 out of 1,000
1.3	2 out of 1,000	20 out of 1,000
0.4	>1 out of 1,000	6 out of 1,000

Estimated risk of lung cancer death as a result of lifetime radon exposure. Radon levels are measured in picocuries per liter (pCi/L). The Enviromental Protection Agency considers a radon level above 4 pCi/Liter to be unsafe. For information about testing your home for radon and what to do if the radon level is high, visit the EPA's Radon Information Site at www.epa.gov/radon.

George Perry Study: AMY-1 Copies among Cultural Groups

The human *AMY-1* gene encodes salivary amylase, an enzyme that breaks down starch. The number of copies of this gene varies, and people who have more copies generally make more enzyme. In addition, the average number of *AMY-1* copies differs among cultural groups.

George Perry and his colleagues hypothesized that duplications of the *AMY-1* gene would confer a selective advantage in cultures in which starch is a large part of the diet. To test this hypothesis, the scientists compared the number of copies of the *AMY-1* gene among members of seven cultural groups that differed in their traditional diets. The graph to the right shows their results.

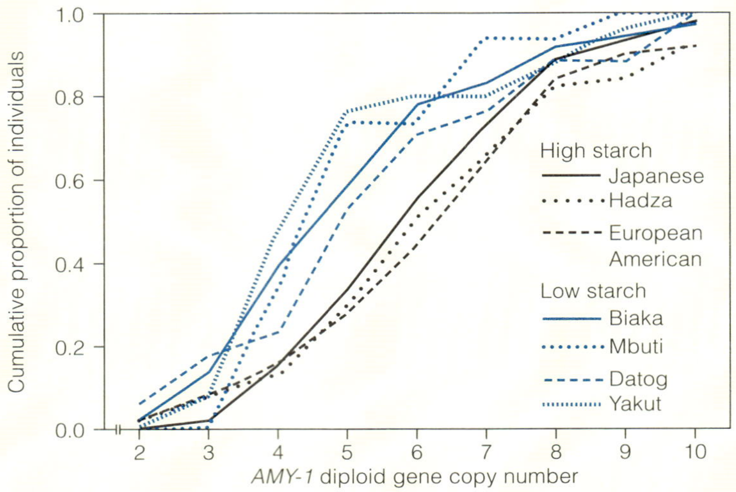

Number of copies of the *AMY-1* gene among members of cultures with traditional high-starch or low-starch diets. The Hadza, Biaka, Mbuti, and Datog are tribes in Africa. The Yakut live in Siberia.

1. Starchy tubers are a mainstay of Hadza hunter–gatherers in Africa, whereas fishing sustains Siberia's Yakut. Almost 60 percent of Yakut had fewer than 5 copies of the *AMY1* gene. What percent of the Hadza had fewer than 5 copies?

2. None of the Mbuti (rain-forest hunter–gathers) had more than 10 copies of *AMY-1*. Did any European Americans?

3. Do these data support the hypothesis that a starchy diet favors duplications of the *AMY-1* gene?

Chensheng Lu Study: Effect of Organic Food on Pesticide Residue

Products labeled as "organic" fill an increasing amount of space on supermarket shelves. What does this label mean? A food that carries the USDA's organic label must be produced without pesticides such as malathion and chlorpyrifos, which conventional farmers typically use on fruits, vegetables, and many grains.

Does eating organic food significantly affect the level of pesticide residues in a child's body? Chensheng Lu of Emory University used urine testing to find out. For fifteen days, the urine of twenty-three children (aged 3 to 11) was monitored for breakdown products of pesticides. During the first five days, children ate their standard, nonorganic diet. For next five days, they ate organic versions of the same types of foods and drinks. Then, for the final five days, the children returned to their nonorganic diet.

		Malathion Metabolite		Chlorpyrifos Metabolite	
Study Phase	No. of Samples	Mean (μg/liter)	Maximum (μg/liter)	Mean (μg/liter)	Maximum (μg/liter)
1. Nonorganic	87	2.9	96.5	7.2	31.1
2. Organic	116	0.3	7.4	1.7	17.1
3. Nonorganic	156	4.4	263.1	5.8	25.3

Above, levels of metabolites (breakdown products) of malathion and chlorpyrifos detected in the urine of children taking part in a study of the effects of an organic diet. The difference in the mean level of metabolites in the organic and inorganic phases of the study was statistically significant. *Right*, the USDA organic food label.

1. During which phase of the experiment did the children's urine contain the lowest level of the malathion metabolite?

2. During which phase of the experiment was the maximum level of the chlorpyrifos metabolite detected?

3. Did switching to an organic diet lower the amount of pesticide residues excreted by the children?

4. Even in the nonorganic phases of this experiment, the highest pesticide metabolite levels detected were far below those known to be harmful. Given this data, would you spend more to buy organic foods?

Drug Treatment to Prevent Masculinization in Females

Adrenal glands normally make a little testosterone, but a mutation in the gene for the enzyme 21-hydroxylase causes excess production of this hormone. A female child who has a 21-hydroxylase deficiency is exposed to abnormally high levels of testosterone during development. This hormone can enlarge her clitoris and cause her labia to fuse, giving her genitals a more male appearance.

The drug dexamethasone slows the adrenal glands' testosterone production. The chart shows data from a study in which doctors gave this drug to pregnant women carrying daughters with 21-hydroxylase deficiency. Sixteen of these women had previously given birth to a daughter with 21-hydroxylase deficiency. These daughters (sisters of the treated newborns) serve as a point of comparison.

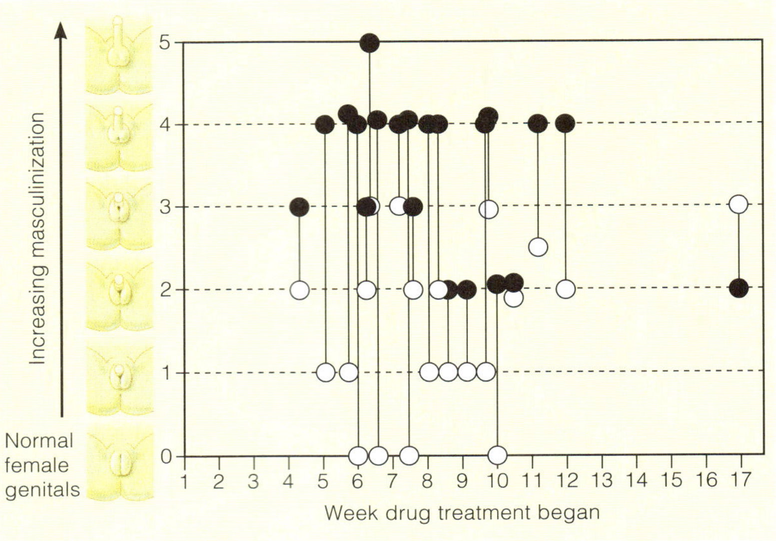

Degree of masculinization of 21-hydroxylase deficient females exposed to dexamethasone in the womb (*open* circles), compared to that of older affected sisters who were untreated during development (*dark* circles). Graphics along the side depict appearance of the newborn's genital area.

1. How many daughters produced by dexamethasone-treated pregnancies had normal female genitals?

2. How many phenotypically normal girls had the women's earlier untreated pregnancies produced?

3. How many women who previously had girls with level 4 or 5 masculinization saw an improvement with treatment?

4. Do the data support the hypothesis that giving the drug-dexamethasone to a pregnant woman can reduce the effects of her developing daughter's 21-hydrolase deficiency?

Multiple Births and Birth Defects

People considering fertility treatments should be aware that such treatments raise the risk of multiple births, and that multiple pregnancies are associated with an increased risk of some birth defects.

The chart shows the results of Yiwei Tang's study of birth defects reported in Florida from 1996 to 2000. Tang compared the incidence of various defects among single and multiple births. She calculated the relative risk for each type of defect based on type of birth, and corrected for other differences that might increase risk such as maternal age, income, race, and medical care during pregnancy. A relative risk of less than 1 means that multiple births pose less risk of that defect occurring. A relative risk greater than 1 means multiples are more likely to have a defect.

1. What was the most common type of birth defect in the single-birth group?

2. Was that type of defect more or less common among the multiple-birth newborns than among single births?

3. Tang found that multiples have more than twice the risk of single newborns for one type of defect. Which type?

4. Does a multiple pregnancy increase the relative risk of chromosomal defects in offspring?

	Prevalence of Defect		Relative
	Multiples	Singles	Risk
Total birth defects	358.50	250.54	1.46
Central nervous system defects	40.75	18.89	2.23
Chromosomal defects	15.51	14.20	0.93
Gastrointestinal defects	28.13	23.44	1.27
Genital/urinary defects	72.85	58.16	1.31
Heart defects	189.71	113.89	1.65
Musculoskeletal defects	20.92	25.87	0.92
Fetal alcohol syndrome	4.33	3.63	1.03
Oral defects	19.84	15.48	1.29

Prevalence, per 10,000 live births, of various types of birth defects among multiple and single births. Relative risk for each defect is given after researchers adjusted for the mother's age, race, previous adverse pregnancy experience, education, Medicaid participation during pregnancy, as well as the infant's sex and number of siblings.

Spread of Africanized Honeybees in the United States

Honeybees disperse by forming new colonies. An old queen leaves the hive along with a group of workers. These bees fly off, find a new nest site, and set up a new hive. Meanwhile, at the old hive, a new queen emerges, mates, and takes over. A new hive can be several kilometers from the old one.

Africanized honeybees form new colonies more often than European ones, a trait that contributes to their spread. Africanized bees also spread by taking over existing hives of European bees. In addition, in areas where European and Africanized hives coexist, European queens are more likely to mate with Africanized males, thus introducing Africanized traits into the colony. The figure to the right shows the counties in the United States in which Africanized honeybees became established from 1990 through 2006.

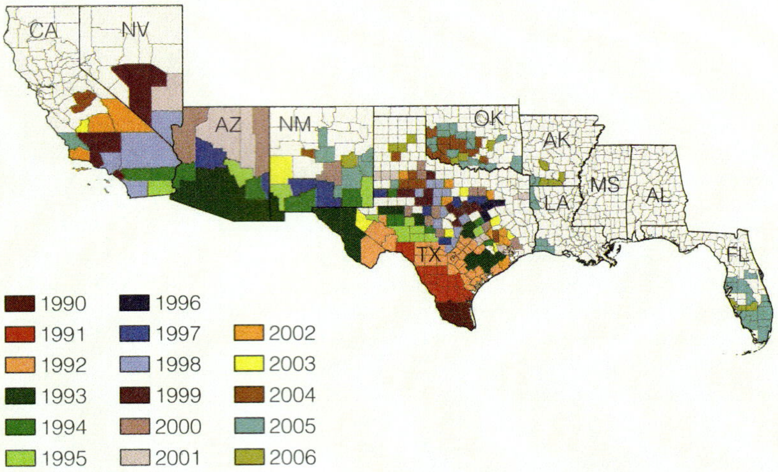

The spread of Africanized honeybees in the United States, from 1990 through 2006. The USDA adds a county to this map only when the state officially declares bees in that county Africanized. Bees can be identified as Africanized on the basis of morphological traits or analysis of their DNA.

1. Where in the United States did Africanized bees first become established?

2. In what states did Africanized bees first appear in 2005?

3. Why is it likely that human transport of bees contributed to the spread of Africanized honeybees to Florida?

4. Based on this map, would you expect Africanized honeybees to colonize additional states in the next five years?

Martin Wikelski Study: Marine Iguana Populations in the Galapagos

In 1989, Martin Wikelski started a long-term study of marine iguana populations in the Galápagos Islands. He marked the iguanas on two islands— Genovesa and Santa Fe—and collected data on how their body size, survival, and reproductive rates varied over time. The iguanas eat algae and have no predators, so deaths are usually the result of food shortages, disease, or old age. His studies showed that numbers decline during El Niño events, when the surrounding waters heat up.

In January 2001, an oil tanker ran aground and leaked a small amount of oil into the waters near Santa Fe—the bar graph to the right shows the number of marked iguanas that Wikelski and his team counted in their census of study populations just before the spill and about a year later.

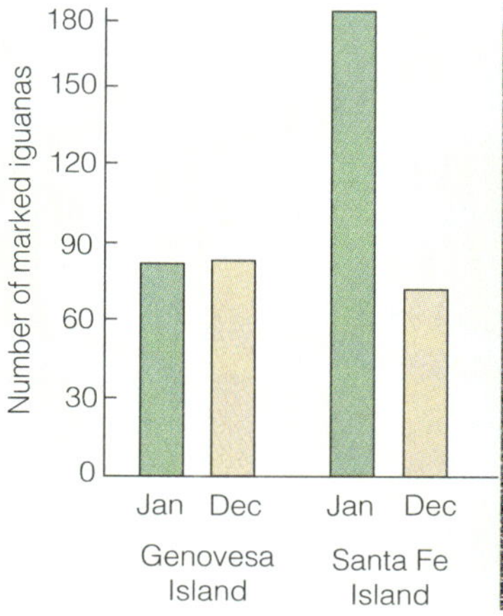

Shifting numbers of marked marine iguanas on two Galápagos islands. An oil spill occurred near Santa Fe just before the January 2001 census (*green* bars). A second census was carried out in December 2001 (*tan* bars).

1. Which island had more marked iguanas at the time of the first census?

2. How much did the population size on each island change between the first and second census?

3. Wikelski concluded that changes on Santa Fe were the result of the oil spill, rather than sea temperature or other climate factors common to both islands. How would the census numbers be different from those he observed if an adverse event had affected both islands?

Methods of Controlling Fire Ant Populations

Ant-decapitating phorid flies are just one of the biological control agents used to battle imported fire ants. Researchers have also enlisted the help of *Thelohania solenopsae*, another natural enemy of the ants. This microsporidian is a parasite that infects ants and shrinks the ovaries of the colony's egg-producing female (the queen). As a result, a colony dwindles in numbers and eventually dies out.

Are these biological controls useful against imported fire ants? To find out, USDA scientists treated infested areas with either traditional pesticides or pesticides plus biological controls (both flies and the parasite). The scientists left some plots untreated as controls. The graph shows the results.

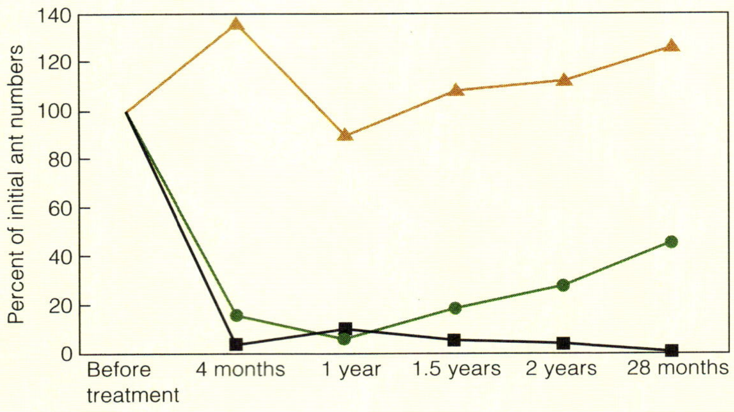

Effects of two methods of controlling red imported fire ants. The graph shows the numbers of red imported fire ants over a 28-month period. *Orange* triangles represent untreated control plots. *Green* circles are plots treated with pesticides alone. *Black* squares are plots treated with pesticide and biological control agents (phorid flies and a microsporidian parasite).

1. How did population size in the control plots change during the first four months of the study?

2. How did population size in the two types of treated plots change during this same interval?

3. If this study had ended after the first year, would you conclude that biological controls had a major effect?

4. How did the two types of treatment (pesticide alone versus pesticide plus biological controls) differ in their longer-term effects?

Changes in Atmospheric Carbon Dioxide Levels

To assess the impact of human activity on the carbon dioxide level in Earth's atmosphere, it helps to take a long view. One useful data set comes from deep core samples of Antarctic ice. The oldest ice core that has been fully analyzed dates back a bit more than 400,000 years. Air bubbles trapped in the ice provide information about the gas content in Earth's atmosphere at the time the ice formed. Combining ice core data with more recent direct measurements of atmospheric carbon dioxide—as in the graphs to the right—can help scientists put current changes in the atmospheric carbon dioxide into historical perspective.

1. What was the highest carbon dioxide level between 400,000 B.C. and 0 A.D.?

2. During this period, how many times did carbon dioxide reach a level comparable to that measured in 1980?

3. The industrial revolution occurred around 1800. What was the trend in carbon dioxide level in the 800 years prior to this event? What about in the 175 years after it?

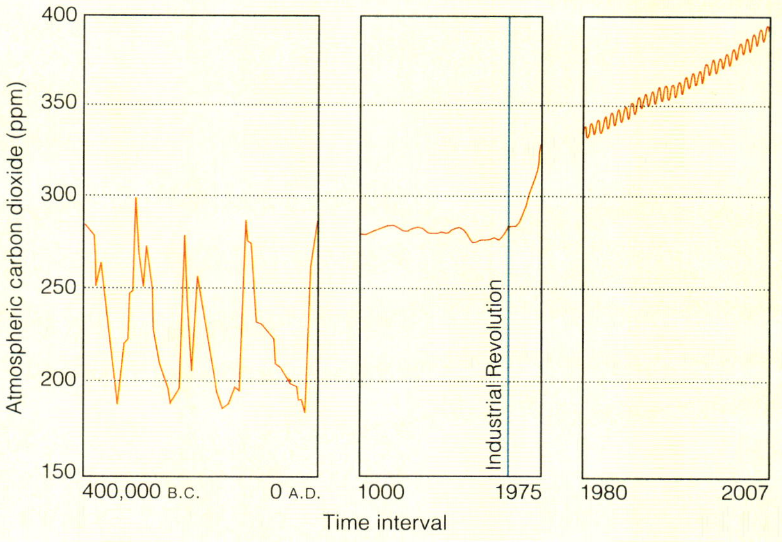

Changes in atmospheric carbon dioxide levels (in parts per million). Direct measurements began in 1980. Earlier data are based on ice cores.

4. Was the rise in the carbon dioxide level between 1800 and 1975 larger or smaller than the rise between 1980 and 2007?

Sea Surface Temperature Anomalies (El Niño and La Niña)

To try to predict the effect of El Niño or La Niña events in the near future, the National Oceanographic and Atmospheric Administration collects information about sea surface temperature (SST) and atmospheric conditions. They compare monthly temperature averages in the eastern equatorial Pacific Ocean to historical data and calculate the difference (the degree of anomaly) to determine if El Niño conditions, La Niña conditions, or neutral conditions are developing. El Niño is a rise in the average SST above 0.5°C. A decline of the same amount is La Niña. The graph shows data for nearly 39 years.

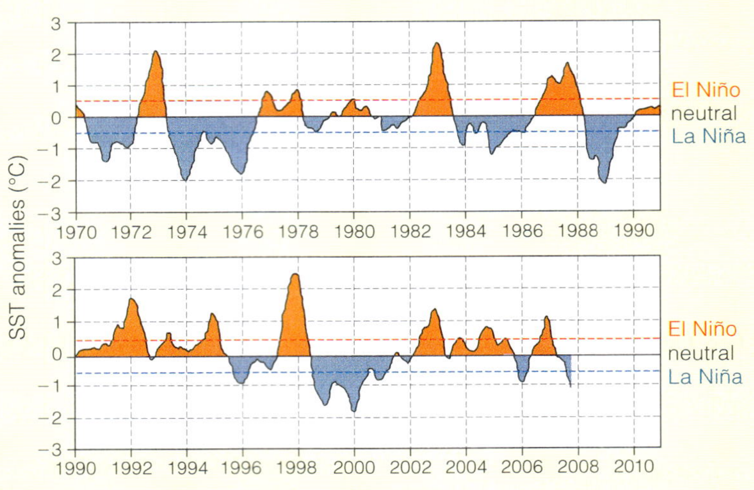

Sea surface temperature anomalies (differences from the historical mean) in the eastern equatorial Pacific Ocean. A rise above the dotted *red* line is an El Niño event, a decline below the *blue* line is La Niña.

1. When did the greatest positive temperature deviation occur during this time period?

2. What type of event, if any, occurred during the winter of 1982–1983? What about the winter of 2001–2002?

3. During a La Niña event, less rain than normal falls in the American West and Southwest. In the time interval shown, what was the longest interval without a La Niña event?

4. What type of conditions were in effect in the fall of 2007 when California suffered severe wildfires?

Biological Magnification in the Arctic

Winds carry chemical contaminants produced and released at temperate latitudes to the Arctic, where the chemicals enter food webs. By the process of biological magnification, top carnivores in arctic food webs—such as polar bears and people—end up with high doses of these chemicals. For example, indigenous arctic people who eat a lot of local wildlife tend to have unusually high levels of polychlorinated biphenyls, or PCBs, in their bodies. The Arctic Monitoring and Assessment Programme studies the effects of these chemicals on health and reproduction. The graph shows the effect of PCBs on the sex ratio at birth in indigenous populations in the Russian Arctic.

1. Which sex is was most common in offspring of women with less than one microgram per millileter of PCB in serum?

2. At what PCB concentrations were women more likely to have daughters?

3. In some Greenland villages, nearly all recent newborns are female. Would you expect PCB levels in those villages to be above or under 4 micrograms per milliliter?

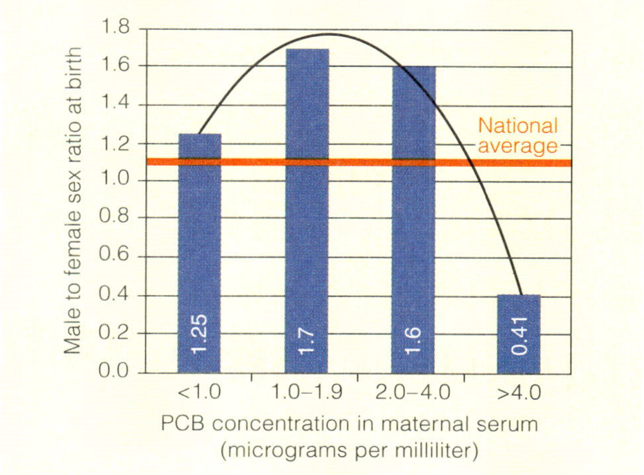

Effect of maternal PCB concentration on sex ratio of newborns in indigenous populations in the Russian Arctic. The red line indicates the average sex ratio for births in Russia—1.06 males per female.